Luis Fernando Aguas Bucheli

Unity and Mugen Editor: Video Game Programming from Scratch

Luis Fernando Aguas Bucheli

Unity and Mugen Editor: Video Game Programming from Scratch

Build virtual worlds and master game creation with Unity and Mugen Editor

ScienciaScripts

Imprint
Any brand names and product names mentioned in this book are subject to trademark, brand or patent protection and are trademarks or registered trademarks of their respective holders. The use of brand names, product names, common names, trade names, product descriptions etc. even without a particular marking in this work is in no way to be construed to mean that such names may be regarded as unrestricted in respect of trademark and brand protection legislation and could thus be used by anyone.

Cover image: www.ingimage.com

This book is a translation from the original published under ISBN 978-613-9-40742-2.

Publisher:
Sciencia Scripts
is a trademark of
Dodo Books Indian Ocean Ltd. and OmniScriptum S.R.L publishing group

120 High Road, East Finchley, London, N2 9ED, United Kingdom
Str. Armeneasca 28/1, office 1, Chisinau MD-2012, Republic of Moldova, Europe
Printed at: see last page
ISBN: 978-620-8-24642-6

Contents

INTRODUCTION

In a world of constantly evolving technology, video game development has become one of the most dynamic and creative areas of software. Video games not only represent a form of entertainment, but also act as a platform for innovation, interactive storytelling and digital art. In this context, Unity and C# have established themselves as fundamental tools for both novice and experienced developers.

Unity, a widely used game engine, provides a robust and flexible platform for creating interactive 2D and 3D experiences. With its intuitive interface and powerful feature set, Unity allows developers to bring their ideas to life, from simple mobile games to complex virtual reality applications. On the other hand, C# is the primary programming language used in Unity, known for its simplicity and efficiency, making it accessible for beginners and powerful enough for advanced developers.

This book, created by Luis Fernando Aguas, founder of Aguaszoft, a company with a deep passion for technology and education, is designed to guide you through the exciting journey of learning Unity and C#. At Aguaszoft, we believe that education is the key to unlocking the creative potential of every individual. Therefore, we have developed this resource with the aim of providing structured and understandable learning, allowing readers to not only acquire technical knowledge, but also get inspired and motivated to create their own projects.

Throughout the pages of this book, we will explore the fundamentals of Unity and C#, covering everything from basic concepts to advanced techniques. We'll start with an introduction to the Unity interface and its main components, and then delve into programming with C#. You will learn how to create and manage objects in Unity, how to apply physics and animations, and how to develop robust and efficient game logic. In addition, we will cover important topics such as performance optimisation, implementing user interfaces, and publishing your game on different platforms.

The approach of this book is practical and project-oriented. We believe that the best way to learn is by doing, so each chapter includes examples and exercises that will allow you to put what you learn into practice. In addition, we have included complete projects at the end of each section, which will challenge you to apply your knowledge and skills to create functional and engaging games.

Unity and C# are not only powerful tools for game development, but also open doors to a wide range of applications in other fields, such as simulation,

architectural visualization, augmented reality, and education. With the skills you will acquire through this book, you will be prepared to explore these opportunities and take your ideas to the next level.
At Aguaszoft, we are committed to your success and growth as a developer. This book is just the beginning of your journey. As you continue to learn, we encourage you to join our developer community, where you can share your projects, receive feedback, and collaborate with other Unity and C# enthusiasts. Together, we can create innovative and exciting interactive experiences that capture the imagination of gamers and users around the world.
So get ready to dive into the fascinating world of game development with Unity and C#. With the gwa and resources provided in this book, you will have all the tools you need to transform your ideas into reality. Welcome to the Aguaszoft community and the beginning of an exciting adventure in game development!

CHAPTER 1

Installing Unity

1.1 Introduction

Unity is one of the world's most popular game development platforms. It allows developers to create games for a variety of platforms, including consoles, PCs, mobile devices and virtual reality. In this chapter, we'll walk you through the process of installing Unity, making sure your development environment is properly configured to start creating your own interactive games and applications.

1.2 System requirements

Before installing Unity, make sure your system meets the minimum requirements:

1.2.1 Requirements for Windows

- **Operating System**: Windows 7 SP1+, Windows 8, Windows 10, and Windows 11 (64-bit)
- **Processor**: CPU with support for SSE2 instructions
- **RAM**: 4 GB RAM (8 GB recommended)
- **Graphics**: DirectX 10 compatible graphics card or higher
- **Disk space**: Minimum 5 GB of available space

1.2.2 Requirements for macOS

- **Operating system**: macOS 10.12 Sierra and later versions
- **Processor**: CPU with support for SSE2 instructions
- **RAM**: 4 GB RAM (8 GB recommended)
- **Graphics**: Metal compatible graphics card
- **Disk space**: Minimum 5 GB of available space

1.2.3 Linux requirements

- **Operating system**: Official supported distributions: Ubuntu 16.04, 18.04, 20.04 (64-bit)
- **Processor**: CPU with support for SSE2 instructions
- **RAM**: 4 GB RAM (8 GB recommended)
- **Graphics**: OpenGL 3.2 or higher compatible graphics card
- **Disk space**: Minimum 5 GB of available space

1.3 Download Unity Hub

The first step to install Unity is to download Unity Hub, an application that manages Unity versions, projects and development environment settings.

1. Visit the official Unity website: Unity3d.com.
2. Click on the "Get Started" button.

3. Select the plan that best suits your needs (you can start with the Personal plan, which is free for small projects).
4. Download the Unity Hub for your operating system (Windows, macOS, Linux).

1.4 Installing the Unity Hub

1.4.1 Installation under Windows

1. Open the downloaded file (UnityHubSetup.exe).
2. Follow the installer's instructions.
3. Once the installation is complete, open Unity Hub.

1.4.2 Installation on macOS

1. Open the downloaded file (UnityHub.dmg).
2. Drag the Unity Hub icon to the Applications folder.
3. Open Unity Hub from the Applications folder.

1.4.3 Linux Installation

1. Open a terminal.
2. Navigate to the directory where the Unity Hub file was downloaded.
3. Run the following command to install Unity Hub:

```
sudo dpkg -i unityhub*.deb
```

4. Open Unity Hub from your applications menu.

1.5 Initial configuration of Unity Hub

When you open Unity Hub for the first time, you will need to make some initial configurations.

1.5.1 Create a Unity Account

1. Open Unity Hub.
2. Click on "Sign In".
3. If you do not have an account, click on "Create One".
4. Follow the instructions to create a Unity account.

1.5.2 Logging in to Unity Hub

1. Open Unity Hub.
2. Click on "Sign In".
3. Enter your credentials and click on "Sign In".

1.6 Installing Unity Editor

Once you have set up Unity Hub, it is time to install Unity Editor, which is the main application you will use to develop your projects.

1.6.1 Selecting a Unity Version

1. Open Unity Hub.
2. Click on the "Installs" tab.

3. Click on "Add".
4. Select the version of Unity you wish to install (it is recommended to select the latest LTS version, which is the version with long-term support).
5. Click "Next".

1.6.2 Selection of Additional Modules

1. After selecting the Unity version, you can choose additional modules according to your needs (e.g. support for Android, iOS, WebGL, etc.).
2. Select the modules you need and click "Next".

1.6.3 Acceptance of Terms and Conditions

1. Check Unity's terms and conditions.
2. Check the acceptance box and click "Done".

1.6.4 Installation

1. Unity Hub will start downloading and installing Unity Editor along with the selected modules.
2. Wait for the installation to complete. This may take some time depending on the speed of your Internet connection.

1.7 Creating your First Unity Project

With Unity installed, you are ready to create your first project.

1.7.1 Creating a New Project

1. Open Unity Hub.
2. Click on the "Projects" tab.
3. Click on "New".
4. Enter a name for your project.
5. Select the project template (e.g. 2D, 3D, HDRP, URP, etc.).
6. Select the location where you want to save your project.
7. Click on "Create".

1.7.2 Exploring the Unity Interface

1. When you open your new project, you will be presented with the Unity Editor interface.
2. Spend some time familiarising yourself with the main panels: Hierarchy, Scene, Game, Inspector and Project.

CHAPTER 2

Unity Basics

Introduction

Unity is one of the world's most popular and versatile game development platforms. In this chapter, we will explore the basics of Unity, covering everything from the user interface to creating a simple project. These fundamentals will provide you with a solid foundation for developing your own games and interactive applications.

2.1.Unity User Interface

2.1.1. Project Window

The Project window displays all the files and assets in your project. This is where you can organise and access your assets, such as 3D models, textures, scripts, and sounds.

2.1.2. Jerarqwa Window

The Jerarqwa window displays all the objects in your current scene. The objects are arranged in a tree structure that reflects the parent-child relationship between them.

2.1.3. Scene Window

The Scene window is where you build and edit the game world. You can drag and drop objects into the scene, move, rotate and scale them.

2.1.4. Game Window

The Game window shows a real-time preview of what your game will look like when it runs. This is where you can test and adjust gameplay.

2.1.5. Inspector Window

The Inspector window displays the properties of the objects selected in the Hierarchy or Project window. Here you can modify components and parameters to adjust the behaviour and appearance of the objects.

1.1.1. Creation of a New Project

1.1.2. Launch Unity Hub

To begin, open Unity Hub, which is an application that makes it easy to manage Unity projects and versions. Click on "New Project" to create a new project.

1.1.3. Template Selection

Unity offers several templates for different types of projects, such as 2D games, 3D games, virtual reality games and more. Select the template that best suits your needs. For this chapter, we will choose the 3D template.

1.1.4. Project configuration

Assign a name to your project and select the location where it will be saved.

Then, click on "Create Project" to start the Unity editor.

2.3.Creation and Manipulation of Objects

2.3.1. Creation of Objects

To create an object, right click on the Jerarqwa window and select "Create Empty" or choose one of the predefined objects such as cubes, spheres, or capsules.

2.3.2. Basic Transformations

Objects in Unity have Transform components that allow you to move, rotate and scale objects. Use the toolbar tools (Move, Rotate, Scale) to adjust objects in the Scene window.

2.4.Components and Scripts

2.4.1. Components

Components are the building blocks that define the behaviour and appearance of objects. All objects have at least one Transform component, but can have many others such as Rigidbody for physics, Collider for collisions, and more.

2.4.2. Scripts

Scripts in Unity are written in C# and are used to add custom behaviour to objects. To create a script, right click in the Project window, select "Create > C# Script", name your script and then open it in the code editor.

2.5.Physics and Collisions System

2.5.1. Components of F^sica

Unity provides components such as Rigidbody to apply physics to objects. A Rigidbody allows an object to move realistically under the influence of forces and collisions.

2.5.2. Colliders

Colliders are components that define the shape of an object for collision purposes. They can be simple shapes such as boxes and spheres, or more complex custom shapes.

2.6.Light and Shadow System

2.6.1. Types of lights

Unity offers various types of lights, such as directional, spot and spotlights, each with their own properties and uses.

2.6.2. Shadow Settings

Lights in Unity can cast shadows, adding realism to the scene. Shadows can be adjusted in the light properties to control their quality and appearance.

2.7.Audio System

2.7.1. Audio Components

Unity provides components such as Audio Source for playing sounds and music. An Audio Source can be added to any object in the scene.

2.7.2. Importation and Use of Sounds

To use sounds in your project, simply drag audio files into the Project window. You can then assign these files to an Audio Source and play them at appropriate times in the game.

CHAPTER 3

Creating and Manipulating Objects in Unity with C#

3.1 Introduction

In this chapter, we will learn how to create and manipulate objects in Unity using the C# programming language. We will see how to instantiate objects in the game world, apply transformations and add custom behaviours through scripting.

3.2 Creating Objects in Unity

3.2.1 Instantiation of Prefabs

In Unity, a prefab is a template of an object that can be reused in-game. Prefabs allow you to create exact copies of an object and are especially useful for generating multiple instances of an enemy, environmental elements, or any other object in the game.

```
using UnityEngine;
public class ObjectSpawner : MonoBehaviour
{
public GameObject prefab;
public Transform spawnPoint;
void Start()
{
Instantiate(prefab, spawnPoint.position, spawnPoint.rotation);
}
}
```

In this example, ObjectSpawner is a script that is attached to an object in the scene. The script instantiates a prefab at a spawn point specified at the start of the game.

3.3 Object Handling

3.3.1 Basic Transformations

Every object in Unity has a Transform component that stores its position, rotation and scale. We can modify these values to move, rotate and scale objects in the game world.

```
using UnityEngine;
public class ObjectMover : MonoBehaviour
{
public float moveSpeed = 5f;
public float rotateSpeed = 50f;
void Update()
```

```
{
// Forward and backward movement
float move = Input.GetAxis("Vertical") * moveSpeed * Time.deltaTime;
transform.Translate(0, 0, move);
// Rotation
float rotate = Input.GetAxis("Horizontal") * rotateSpeed * Time.deltaTime;
transform.Rotate(0, rotate, 0);
}
}
```

This script allows you to move an object forwards and backwards using the arrow keys or W and S, and to rotate it left and right using the arrow keys or A and D.

3.4 Interaction between Objects

3.4.1 Collisions and Triggers

To detect collisions between objects, we need to add Collider components to the objects and use C# functions to handle the interactions.

```
using UnityEngine;
public class CollisionDetector : MonoBehaviour
{
void OnCollisionEnter(Collision collision)
{
Debug.Log("Collision detected with " + collision.gameObject.name);
}
void OnTriggerEnter(Collider other)
{
Debug.Log("Trigger detected with " + other.gameObject.name);
}
}
```

- OnCollisionEnter is called when the object collides with another object that also has a Collider and a Rigidbody.
- OnTriggerEnter is called when an object enters a designated Trigger area.

3.5 Adding Personalised Behaviour

We can add custom scripts to our objects so that they have specific behaviours. For example, a script that makes an enemy chase the player.

```
using UnityEngine;
public class EnemyAI : MonoBehaviour
{
```

```
public Transform player;
public float speed = 2f;
void Update()
{
Vector3 direction = player.position - transform.position;
direction.Normalize();
transform.position += direction * speed * Time.deltaTime;
}
}
```

In this example, the enemy will continuously move towards the player's position.

CHAPTER 4

Master Scripted Programming in Unity

4.1 Introduction to Unity Programming

Unity is a powerful platform for the development of 2D and 3D interactive games and applications. Programming in Unity is mainly done using the C# programming language. This chapter will guide you through the basic and advanced concepts of scripting in Unity, providing detailed and practical examples.

4.2 Development Environment Configuration

Before you start programming in Unity, make sure you have Unity Hub and the latest version of the Unity Editor installed. You will also need an IDE such as Visual Studio or Visual Studio Code to edit your scripts.

1. **Install Unity Hub and Unity Editor**: Download and install Unity Hub from the official Unity website. Through Unity Hub, install the Unity Editor version of your choice.
2. **Install Visual Studio**: During the installation of Unity, you will be given the option to install Visual Studio, which is highly recommended for seamless integration.

4.3 Script Structure in Unity

A script in Unity is a code file that inherits from the MonoBehaviour class. Here is a basic example of a C# script:

```
using UnityEngine;
public class MyPrimerScript : MonoBehaviour
{
// This method is called once when starting the game void Start()
{
Debug.Log("Hello, Unity!");
}
// This method is called once per frame
void Update() {
}
}
```

- Start(): Called once when the script is executed for the first time.
- Update(): Called once per frame and is ideal for handling logic that needs to be constantly checked.

4.4 Creation and Linking of Scripts to Objects

1. **Create a New Script**:

◦ In the Project Panel, right-click and select Create > C# Script.
◦ Name your script appropriately (e.g., PlayerController).

2. **Bind the Script to an Object**:

◦ Drag and drop the script on the object you want it to run on from the Project Panel to the object in the hierarchy.
Alternatively, select the object in the hierarchy and in the Inspector Panel, click on Add Component and select your script.

4.5 Common MonoBehaviour Methods

Unity provides several MonoBehaviour methods that you can override to define the behaviour of your game:

- Awake(): Called before Start(), ideal for initialisation that must occur before other scripts access this object.
- FixedUpdate(): Called at fixed intervals, ideal for physics.
- LateUpdate(): Called after Update(), useful for tracking objects that may have moved in Update().
- OnCollisionEnter(), OnTriggerEnter(): Methods to detect collisions and triggers.

4.6 Manipulation of Components and GameObjects

A key aspect of programming in Unity is the manipulation of components and GameObjects. Here are some common examples:

1. **Access to Components**:

```
void Start()
{
// Accessing the Rigidbody component of the object
Rigidbody rb = GetComponent<Rigidbody>();
rb.AddForce(Vector3.up * 10);
}
```

2. **Instantiate and Destroy Objects**:

```
public GameObject prefab;
void Update()
{
if (Input.GetKeyDown(KeyCode.Space))
{
// Instantiate a prefab object
Instantiate(prefab, transform.position, transform.rotation);
}
}
```

3. **Destroy Objects**:

```
void OnCollisionEnter(Collision collision)
{
if (collision.gameObject.tag == "Enemy")
{
Destroy(collision.gameObject);
}
}
```

4.7 Coroutines

Coroutines allow you to execute methods that can pause their execution and return to it in the next frame, creating timed behaviours.

```
void Start()
{
StartCoroutine(ExampleCoroutine());
}
IEnumerator ExampleCoroutine()
{
Debug.Log("Start coroutine");
yield return new WaitForSeconds(2);
Debug.Log("2 seconds later");
}
```

4.8 Advanced Practices

1. **Delegates and Events**: Delegates and events allow you to create notification systems and handle interactions between scripts efficiently.

```
public delegate void MyDelegate();
public static event MyDelegate OnEvent;
void Start()
{
OnEvent += ResponseMethod;
OnEvent?.Invoke();
}
void MethodResponse()
{
Debug.Log("Event received");
}
```

2. **ScriptableObjects**: ScriptableObjects are an efficient way to store data and create reusable objects.

```
[CreateAssetMenu(fileName = "PlayerData", menuName = "PlayerData")]
public class PlayerData : ScriptableObject
{
public string name;
public int health;
}
```

4.9 Debugging and Optimisation

1. **Debug Debugging**: Use Debug.Log, Debug.Warning, and Debug.Error to debug and trace problems in your code.

```
Debug.Log("Debug message");
Debug.LogWarning("Warning");
Debug.LogError("Error");
```

2. **Optimisation**:

- Minimise the use of Update() where possible.
- Use Object Pooling to handle frequent object instantiation and destruction.
- Optimises the use of physics and collisions by properly configuring the collision components.

CHAPTER 5

Physics and Collisions in Unity 2D

In this chapter, we will explore the fundamental concepts of physics and collisions in Unity 2D. You will learn how to use Unity's physics system to bring your objects to life in a 2D environment and how to handle collisions between them.

5.1 Introduction to Unity 2D logic

Unity provides a robust physics engine to simulate the realistic behaviour of objects in a 2D environment. This includes gravity, forces and collisions.

5.1.1 Physics System Configuration

To start using physics in Unity 2D, you must ensure that your objects have Rigidbody2D and Collider2D components. These components allow objects to physically interact with each other^ and respond to forces and collisions.

5.2 Music components

5.2.1 Rigidbody2D

The Rigidbody2D component allows an object to be controlled by the physics engine. This includes applying forces, velocity and other physics properties. To add a Rigidbody2D to an object:

1. Select the object in the hierarchy.
2. Go to the Inspector and click on "Add Component".
3. Search and select "Rigidbody2D".

Key Properties of Rigidbody2D:

- **Mass:** Defines the mass of the object.
- **Gravity Scale:** Controls how much gravity affects the object.
- **Drag:** Controls the resistance to movement.

5.2.2 Collider2D

The Collider2D defines the shape of the object for collision detection. There are several types of colliders you can use, such as Box Collider 2D, Circle Collider 2D and Polygon Collider 2D.

To add a Collider2D:

1. Select the object in the hierarchy.
2. Go to the Inspector and click on "Add Component".
3. Search and select the type of Collider2D you need (e.g. "Box Collider 2D").

5.3 Collision Management

Unity provides several methods to detect and respond to collisions between objects. These methods are implemented in scripts attached to objects that have colliders.

5.3.1 Collision Methods

- **OnCollisionEnter2D:** Called when the object starts to collide with another object.
- **OnCollisionStay2D:** Called every frame while the object is still colliding with another object.
- **OnCollisionExit2D:** Called when the object stops colliding with another object.

```
void OnCollisionEnter2D(Collision2D collision)
{
Debug.Log("Collision detected with: " + collision.gameObject.name);
}
```

5.3.2 Trigger Methods

In addition to normal collisions, Unity also allows the use of "Triggers", which are colliders that do not produce a physical response but can be used to detect interactions.

- **OnTriggerEnter2D:** Called when another collider enters the trigger.

OnTriggerStay2D: Called every frame while another collider is inside the trigger.

OnTriggerExit2D: Called when another collider exits the trigger.

```
void OnTriggerEnter2D(Collider2D other)
{
Debug.Log("Trigger enabled by: " + other.gameObject.name);
}
```

5.4 Applying Forces and Movements

Unity allows forces to be applied to objects with Rigidbody2D to simulate more realistic movements.

5.4.1 Apply Force

You can apply force to a Rigidbody2D using the AddForce method.

```
Rigidbody2D rb = GetComponent<Rigidbody2D>();
rb.AddForce(new Vector2(10f, 0f), ForceMode2D.Impulse);
```

5.4.2 Apply Speed

You can also directly change the speed of a Rigidbody2D.

```
Rigidbody2D rb = GetComponent<Rigidbody2D>();
rb.velocity = new Vector2(5f, 0f);
```

5.5 Practical Example: Creating a Playable Character

We will create a character that can move and jump using Unity 2D physics.

5.5.1 Character Configuration

1. Create a 2D object (e.g. a sprite) and name it "Player".
2. Adds a Rigidbody2D component to the "Player" object.
3. Add a Box Collider2D to define the shape of the collider.

5.5.2 Motion Script

Create a script called PlayerMovement and attach it to the "Player" object.

```
using UnityEngine;
public class PlayerMovement : MonoBehaviour
{
public float moveSpeed = 5f;
public float jumpForce = 10f;
private Rigidbody2D rb;
void Start()
{
rb = GetComponent<Rigidbody2D>();
}
void Update()
{
float moveInput = Input.GetAxis("Horizontal");
rb.velocity = new Vector2(moveInput * moveSpeed, rb.velocity.y);
if (Input.GetButtonDown("Jump") && Mathf.Abs(rb.velocity.y) < 0.01f)
{
rb.AddForce(new Vector2(0f, jumpForce), ForceMode2D.Impulse);
}
}
}
```

CHAPTER 6

3D development in Unity

6.1 Introduction to Unity 3D

6.1.1 What is Unity?

Unity is a cross-platform game development engine used to create interactive 2D and 3D applications. It provides an intuitive development environment and powerful tools for designing, developing and deploying games and applications on various platforms such as PC, consoles, mobile devices and virtual reality.

6.1.2 Unity installation and configuration

To get started with Unity, you first need to download the Unity Hub from the official Unity website (https://unity.com/). The Unity Hub allows you to install different versions of Unity, manage your projects and access learning resources. Once the Unity Hub is installed, you can select the version of Unity you want to install and configure the modules needed for your project.

6.1.3 Unity interface

The Unity interface is composed of several windows and panels that facilitate the development of games and applications. Some of the main windows are:

- **Scene**: This is where you can build and organise your 3D world.
- **Game**: Shows what your game will look like when it runs.
- **Hierarchy**: Displays all objects in the current scene.
- **Project**: Contains all the assets of your project.
- **Inspector**: Displays and allows you to edit the properties of the selected object.

6.2 3D Object Creation

6.2.1 Types of 3D objects in Unity

Unity allows you to create different types of 3D objects, such as primitives (cubes, spheres, cylinders, etc.), imported models and custom objects. Primitives are useful for rapid prototyping and basic scene construction.

6.2.2 Importing 3D models

You can import 3D models created in external software such as Blender, Maya or 3ds Max. To import a model, simply drag it into the Assets folder in Unity. Unity supports common formats such as FBX, OBJ and Collada.

6.2.3 Creation and manipulation of primitives

To create a primitive, go to the GameObject > 3D Object menu and select the type of object you wish to create. Once created, you can move, rotate and scale the object using the transformation tools at the top left of the interface.

6.3 Materials and Textures

6.3.1 Introduction to materials

Materials in Unity determine how the surface of an object will look. A material can have various properties, such as colour, texture and reflectivity.

You can create a new material in the Assets folder and assign it to an object by dragging it to the object's Renderer component.

6.3.2 Application of textures

Textures are images that are applied to the surface of a material to give it more detail. To apply a texture, simply drag it onto the material's albedo field in the Inspector.

6.3.3 Shaders and visual effects

Shaders are small programs that determine how an object is rendered on the screen. Unity provides a variety of default shaders, but you can also create custom shaders using the ShaderLab or HLSL language.

6.4 Lighting

6.4.1 Types of lights in Unity

Unity offers several types of lights to illuminate your 3D scenes:

- **Directional Light**: Simulates sunlight, illuminates the entire scene in a specific direction.
- **Point Light**: Emits light in all directions from a specific point, similar to a light bulb.
- **Spot Light**: Emits light in a cone, useful for spotlights and torches.
- **Area Light**: Emits light from a rectangular surface (only available in advanced rendering).

6.4.2 Global illumination configuration

Global lighting in Unity includes ambient lighting and light mapping. You can adjust the ambient lighting from the Window > Rendering > Lighting menu. Here you can also configure light mapping to improve the performance and visual quality of your scene.

6.4.3 Shadows and lighting effects

Shadows add realism to your scene. You can enable and configure shadows in the light properties. Unity also offers advanced lighting effects such as bloom, volumetric lights and dynamic lights.

6.5 Physics and Collisions

6.5.1 Unity's system of features

Unity has a built-in physics engine that simulates the physical interactions between objects. You can add components such as Rigidbody to make objects respond to gravity and other physical forces.

6.5.2 Collisions and triggers

Collisions are handled using colliders, which are components that define the collision shape of an object. Triggers are colliders that do not physically respond to collisions but can detect when an object enters or leaves them.

6.5.3 Rigidbodies and motion control

Rigidbodies allow objects to be affected by physical forces such as gravity and collisions. You can control the movement of an object through scripts that apply forces or modify rigidbody properties.

6.6 Animations

6.6.1 Animation system in Unity

Unity's animation system allows you to create and control complex animations. Use the Animator Controller to manage and play animations on your objects.

6.6.2 Creation of animations

You can create animations directly in Unity using the Animation window. This tool allows you to record and edit animations frame by frame.

6.6.3 Animation control by scripts

You can control scripted animations using the Animator component. This allows you to play, pause and change animations in response to game events.

6.7 Audio

6.7.1 Importing audio files

To import audio files, simply drag them into the Assets folder. Unity supports common formats such as WAV, MP3 and OGG.

6.7.2 Configuration of audio sources and audio listeners

An AudioSource is a component that plays sounds. An AudioListener is a component that listens to sounds, usually attached to the main camera. You can configure these components to play and control the audio in your scene.

6.7.3 Audio control via scripts

You can use scripts to control the AudioSource, allowing you to play, pause and adjust the audio volume dynamically during the game.

CHAPTER 7

Unity Integration with Game Consoles

In this chapter, we'll explore how Unity makes it easy to integrate games with some of the most popular video game consoles on the market. From PlayStation and Xbox to Nintendo Switch and beyond, Unity offers powerful tools that allow developers to bring their creations to these coveted platforms.

Supported Platforms

Unity supports a wide range of video game consoles, providing developers with direct access to platforms such as PlayStation (PS4, PS5), Xbox (Xbox One, Xbox Series X|S), Nintendo Switch, and other less common but equally important consoles.

SDKs and Documentation

Each console has its own SDK (software development kit) that developers must integrate with Unity. These SDKs allow access to console-specific features, such as user controls, networking functionality, and performance optimisation. Unity simplifies this process by providing detailed documentation and practical examples for each platform.

Performance Optimisation

To take full advantage of the consoles' capabilities, Unity allows developers to optimise their games in terms of memory usage, advanced graphics processing and storage efficiency. This ensures that games can deliver a smooth, high-quality experience on every supported platform.

Certification and Publication

Prior to the release of a game on a console, it is crucial to comply with the certification standards of the respective platform. Unity facilitates this process by providing integrated tools to thoroughly test and resolve potential issues prior to final certification.

Access to Specific Features

Each console has unique features that developers can take advantage of, such as special sensors, virtual reality functionality and streaming capabilities. Unity offers access to these features through its APIs and development tools, allowing creators to innovate and differentiate their games in the marketplace.

CHAPTER 8

Optimisation and Development for Mobile Devices in Unity

In this chapter, we will explore how to optimise and develop games for mobile devices using Unity. Mobile devices present unique challenges due to their hardware limitations and variability in performance capabilities. We will learn key strategies to ensure that our games are efficient and provide an optimal user experience.

Introduction to Mobile Optimisation:

Mobile devices, such as smartphones and tablets, have limited resources compared to desktop and console platforms. It is critical to understand these limitations in order to optimise our games and deliver a smooth and engaging experience. Optimisation not only improves performance, but can also extend battery life and reduce device heating issues.

Project configuration in Unity:

When starting a project in Unity for mobile devices, it is crucial to make specific adjustments. This includes the initial configuration of the project adapted to the characteristics of mobile platforms, such as iOS and Android. Proper platform selection and performance configuration are key steps to optimise the game's performance on different devices.

Resource Optimisation:

Efficient management of resources such as textures, models and animations is essential to maintain optimal performance. Using LOD (Level of Detail) techniques to reduce the polygon count of 3D models and adjusting texture quality through compression are best practices to minimise memory usage without compromising visual quality.

Rendering Optimisation:

To improve rendering efficiency on mobile devices, it is essential to implement techniques such as culling to avoid rendering objects outside the player's field of view. The use of batching, which groups similar objects to reduce the number of rendering steps, and the control of resolution and p^xel density are additional strategies to optimise graphics performance.

Code and Script Optimisation:

Identifying and correcting bottlenecks in the code is crucial to optimise overall game performance. Using object pools to efficiently manage dynamic resources and applying optimised algorithms for physics and game logic are effective practices to improve code efficiency and reduce resource consumption.

Testing and profiling:

The testing process on real mobile devices and simulators is essential to validate the performance and stability of the game. Using Unity's tools for performance profiling and analysis allows you to identify areas for improvement and make adjustments based on concrete metrics obtained during testing.

Adaptation of Controls and UI for mobiles:

Designing intuitive and responsive touch controls is essential to optimise gameplay on mobile devices. Adapting the user interface for different resolutions and aspect ratios, as well as implementing haptic and sound feedback, contribute significantly to improving the user experience on mobile devices.

Final Optimisation and Publication:

Before the release of the game, it is crucial to perform final adjustments and extensive testing to ensure that it meets the expected performance and quality standards. Preparing optimised distribution packages for app shops such as the App Store and Google Play is the last step before releasing the game to the market. In addition, establishing strategies to maintain and update the game after launch is critical to respond to user feedback and continuously improve the experience.

CHAPTER 9

AI Implementation in Unity: Movement and Decision Making

Introduction

In this chapter, we will explore how to integrate artificial intelligence into your Unity games. We will focus on two key aspects: the movement of AI-controlled agents and behaviour-based decision making.

Section 1: Configuration of the environment and actors

1. **Preparing the development environment**
   - Unity configuration for AI development.
     - Open Unity and create a new project or use an existing one.
     - Make sure you have the necessary AI tools installed in Unity, such as the NavMesh navigation system.
     - Import additional packages as needed for specific behaviours.
2. **Creation of AI-controlled agents**
   - Design of characters or entities that will act as agents.
     - Create character models or select available assets.
     - Assign basic components such as Rigidbody for physics and Colliders for collision detection.
     - Set up basic animations if necessary for movement.

Section 2: AI-controlled movement

1. **Implementation of motion systems**
   - Basic movement: linear and rotational movement.
     - Create C# scripts to control the movement of the agent.
     - Implements functions to move the agent in a specific direction using transformations.
     - Manages the rotation of the agent to face it in the correct direction of movement.
   - Use of physics and collisions to avoid obstacles.
     - Integrates colliders and raycasts to detect obstacles in front of the agent.
     - Implements logic to divert movement or avoid collisions.
   - Integration of predefined routes and patrols.
     - Defines waypoints or patrol points for the officer.
     - Develops a system for the agent to follow a predefined route using waypoints.
2. **Navigational intelligence**
   - Use of navigation systems such as NavMesh for efficient movement.
     - Set up a NavMesh in your scene to define navigation areas.

- Assign NavMeshAgent components to your agent to allow it to navigate using NavMesh.
- Schedule behaviours for the agent to move to specific destinations within the NavMesh.

Section 3: Decision-making and behaviours

1. **Introduction to behaviour**

 ○ Definition of basic and complex behaviours.

- Establishes different behavioural states for the agent (patrolling, following, evasion, waiting, etc.).
- Defines transitions between states based on specific conditions (e.g. player detection, collision with obstacles, etc.).

 ○ Use of state machines to manage behavioural states.

- It implements a state machine using patterns such as State or Behaviour Tree.
- It schedules smooth transitions between states so that the agent responds coherently to its environment.

2. **Implementation of decision-making logic**

 ○ Design of algorithms to decide actions based on objectives and conditions.

- Develops decision-making algorithms for the agent to select actions such as following the player, searching for objects, or attacking enemies.
- It uses AI techniques such as A* search to plan routes to specific targets.

 ○ Examples of behaviours: patrolling, searching, evading, tracking.

- It implements specific behaviours using the developed algorithms and logic.
- Test and adjust these behaviours to ensure that they are effective and realistic within the context of the game.

Section 4: Optimisation and debugging

1. **Performance optimisation**

 ○ Resource management and optimisation of AI calculations.

- Minimises resource usage by controlling the AI update frequency.
- Profiles performance to identify and correct bottlenecks.

 ○ Performance testing and code profiling.

- Use Unity tools to evaluate the performance of your AI.
- Conduct extensive testing to ensure that AI works efficiently even under demanding conditions.

2. **Behavioural debugging**

 ○ Use of debugging tools in Unity to track AI decisions.

- Implements logs and visualisations in the user interface to show the internal state of the AI.

- Use Unity's debug mode to inspect variables and behavioural states in real time.

- Correction of common errors and improvements in behavioural logic.
  - Identifies and corrects logic errors in AI behaviours.
  - Itera on behavioural design to improve gameplay and player interaction with AI-controlled agents.

CHAPTER 10

Advances in Futuristic Technology with Unity

Unity has evolved beyond being a simple game development platform. Thanks to revolutionary advances in artificial intelligence and augmented reality, Unity has become the nerve centre of global digital interaction.

Advanced Artificial Intelligence: Unity AI, powered by deep learning algorithms, not only powered characters and virtual environments in games, but was also instrumental in automating complex tasks in everyday life. From personal virtual assistants to AI-assisted medical diagnostic systems, Unity AI has transformed the way we interact with technology.

Ubiquitous Augmented Reality: Unity's integration with augmented reality technologies created a world where the boundary between the digital and the physical blurred. From AR glasses that provided instant contextual information to real-time interactive simulations of urban spaces, Unity AR has redefined the human experience in both virtual and real environments.

Complete Simulation of the Environment: With its ability to handle vast amounts of data and process complex models, Unity has become the platform of choice for the simulation of complex systems. From global climate simulations to urban planning and virtual space exploration, Unity enables scientists and planners to envision and model scenarios with an unprecedented level of detail.

Real-Time Global Collaboration: Unity's advanced network infrastructure enabled real-time global collaboration on projects of massive scale. Teams of researchers, designers and developers could work together virtually as if they were physically present in the same place, facilitating scientific discoveries, technological breakthroughs and artistic creations never before imagined.

CHAPTER 11

Artificial Intelligence Implementation for Ghosts

Objective of the Chapter

In this chapter, we will learn how to implement artificial intelligence (AI) for the ghosts in our Pacman game. We will use classical algorithms to control the movement of the ghosts and improve the gameplay.

Chapter Contents

1. **Revision of the Ghost Design**

○ Before we start with the AI implementation, we will review the current design of the ghosts in our game. We will recall their basic behaviours and how they interact with Pacman.

2. **Introduction to Phantom AI**

○ Artificial intelligence plays a crucial role in modern games, directly influencing the player's experience. In our Pacman game, the AI of the ghosts will determine how they chase Pacman and how they avoid obstacles.

3. **Implementation of Basic Behaviour**

○ We will start by implementing a basic behaviour for the ghosts. At this stage, ghosts can move randomly through the maze without regard to Pacman. This will serve as a base on which we will build more complex behaviours.

```
// Basic random movement implementation for a ghost void MoveRandomly() {
// Logic for moving the ghost randomly through the maze
}
```

4. **Pathfinding Algorithm**

○ We will introduce a path-finding algorithm to allow ghosts to chase Pacman more intelligently. We will use the A* (A-star) algorithm, which is efficient for maze games and allows finding the shortest route between two points.

```
// Basic pathfinding implementation (A*) for ghosts to chase Pacman
void MoveTowardsPacman() {
// Implementation of A* to find the way to Pacman
}
```

5. **Obstacle Management and Movement Decisions**

○ Ghosts must be able to avoid obstacles, such as walls and other ghosts. We will implement logic to make the ghosts make intelligent movement decisions, choosing alternative routes when they encounter an obstacle on their way to Pacman.

```
// Implementation of logic for obstacle avoidance and movement decisions
void AvoidObstaclesAndMakeDecisions() {
```

// Logic to avoid obstacles and make decisions based on the environment }

6. **Tuning and Optimisation**

o We will adjust the AI parameters to balance the difficulty of the game. We will ensure that ghosts are challenging but not impossible for Pacman to evade. We will also optimise AI performance to ensure that the game runs smoothly even on less powerful devices.

7. **Chapter Challenge**

o Implement an AI system for the ghosts that allows them to chase Pacman effectively. Make sure the ghosts can navigate the maze, avoid obstacles and create a challenging experience for the player.

CHAPTER 12

Level Development with the Mugen Editor

Objective of the Chapter

In this chapter, we will explore how to use Mugen's level editor to create custom scenarios and add visual and playable depth to your fighting game. We will learn how to manage layers, import graphical assets and set up interactions to create unique gameplay experiences.

Chapter Contents

1. **Introduction to the Mugen Editor**

○ Mugen's level editor is an essential tool for customising fighting games, allowing developers and designers to create unique scenarios tailored to the aesthetics and mechanics of the game. This chapter will focus on how to effectively use this tool to maximise the visual and playable potential of the game.

2. **Editor Interface**

○ When starting the Mugen editor, we are confronted with an interface divided into key areas:

- **Tool Panels:** Here you will find options for managing layers, adding decorative elements and configuring layer properties.
- **Layers Panel:** Allows you to manage the layers of the level, essential for creating visual depth through the use of backgrounds and foreground elements.
- **Objects Panel:** Where the interactive elements of the level, such as moving platforms, doors or traps, are added and managed.
- **Display Area:** Where the level is previewed as it is being designed.

3. **Layers and Backgrounds Configuration**

○ Layers are essential to create the illusion of depth and movement in the game. The Mugen editor allows you to set up multiple layers with different scrolling speeds (parallax) to simulate movement and create dynamic backgrounds.

- **Add and Manage Layers:** Use options in the layers panel to create new layers, adjust their order and set the scrolling speed.
- **Background Settings:** Import graphic assets for static or animated backgrounds. Adjust the position and movement for each background layer as needed.

4. **Graphics Import and Edition**

○ To maintain the visual coherence of the game, it is crucial to import and edit graphics properly.

- **Sprite Import:** Use import tools to add character sprites, objects and backgrounds to your Mugen project.
- **Editing and Optimisation:** Adjust sprites to match the game's art style and optimise the file size to improve performance.

5. **Configuration of Interactive Objects**

   - Interactive objects enrich the gameplay by providing unique challenges and mechanics for players.

- **Add Interactive Objects:** Use the object panel to select and place items such as moving platforms, switches or traps.
- **Property Settings:** Defines the properties and behaviours of each interactive object, such as its movement, activation by the player or deactivation conditions.

6. **Optimisation and Testing**

   - As you develop the level, it is important to optimise its performance and perform extensive testing to ensure a smooth and error-free gaming experience.

- **Performance Optimisation:** Reduce resource usage by adjusting the number of layers and sprite size as needed.
- **Gameplay Testing:** Performs tests to verify the functionality of interactive objects, player navigation and visual consistency.

7. **Chapter Desatio**

   - To put what you have learned into practice, create a complete level using the Mugen editor. Make sure to implement multiple layers with dynamic backgrounds and at least one interactive element to enhance the gameplay experience.

Introduction:

In this special chapter of Mugen Editor, we explore the nostalgia and influence of classic games on the contemporary creation of fighting game characters and environments. From arcade to retro consoles, many of these games have left an indelible mark on gaming culture.

1. *Street Fighter II: The Fighting Revolution*

- **Inspired Characters:** Creation of characters that pay tribute to iconic Street Fighter II fighters such as Ryu, Chun-Li, and Blanka. Inclusion of signature moves and techniques from Capcom's legendary game.
- **Emblematic Stages:** Stage designs that capture the essence of Street Fighter II's most famous locations, such as Kanzuki Temple and the Sumo Ring.

2. *Mortal Kombat: The Bloody Battle for Supremacy*

- **Fatalities and Brutalities:** Implementation of finishing moves and finishing sequences reminiscent of the famous Fatalities and Brutalities from Mortal Kombat. Graphic and sound details to increase the intensity and excitement of the fights.
- **Tournament Arena:** Creation of a stage inspired by the classic Mortal Kombat tournament stage, with interactive details and deadly traps.

3. *Super Smash Bros: The Legendary Crossover*

- **Cameos and Crossovers:** Incorporation of guest characters from other classic and contemporary games, allowing epic confrontations between konos of the videogame world.
- **Inspired Levels:** Scenario designs that combine elements from multiple game universes, offering unique and visually spectacular challenges.

LABORATORY PRACTICES

Unity is a development platform for video games and interactive experiences in 3D and 2D, widely used in the industry for its versatility and power. It provides developers with a complete suite of tools to create and optimise video games, simulations and virtual and augmented reality applications.

Course objectives

The main objective of these lab practicals is to introduce students to the use of Unity as a tool for the development of video games and interactive applications. At the end of the course, students should be able to:

- Understand the basic concepts of the Unity interface.
- Use the Unity graphics engine to create scenes and characters.
- Implement C# scripts to control the behaviour of game objects.
- Apply basic physics and animation techniques in their projects.
- Develop a complete project from the initial idea to a playable prototype.

Methodolt^a

The course will be conducted through a series of hands-on labs that will allow students to learn by doing. Each practicum will address different aspects of Unity development, from the fundamentals to more advanced techniques. Students will work in teams to encourage collaboration and the exchange of ideas, and each session will include a brief theoretical explanation followed by practical exercises.

Tools and Resources

- **Unity Hub**: Application to manage Unity installations and projects.
- **Unity Editor**: Integrated development environment where projects are built and edited.
- **C#**: Programming language used to create scripts in Unity.
- **Asset Store**: Repository in Knea where you can download resources such as models, textures and scripts to use in your projects.

1. **TOPIC:** Introduction to Unity
2. **OBJECTIVES:**

- To familiarise with Unity as a comprehensive platform for the development of interactive 3D content in real time.
- Analyse the Unity interface.

3. **INTRODUCTION:**

Unity is more than an engine: it is the world's leading platform for creating and operating real-time interactive 3D (RT3D) content. Game programmers, artists, architects, automotive designers and filmmakers, among others, use Unity to bring the products of their imagination to life.

Among the key features of Unity:

- **Comprehensive Platform:** Unity is not only a game engine, but also a complete platform for creating interactive experiences. From video games to virtual reality (VR) and augmented reality (AR) applications, Unity covers it all.
- **Visual Editor:** The Unity Editor provides an intuitive graphical interface for designing scenes, importing assets, adjusting lighting and more. It is a powerful tool for creating virtual worlds.
- **Active Community:** Unity has a global community of creators, where you can share knowledge, get answers to your questions and seek inspiration.
- **Free Learning:** Unity Learn offers over 750 hours of free, on-demand learning content. You can explore specific paths based on your interests, such as Unity basics, VR development and more.
- **Download and Install:** To get started, download and install the Unity Editor. If you are new to Unity, you can use the free version called Unity Personal.

Install unity following this link: https://unity.com/download

After installing, we run, it generates a page to register.

After registering and logging in:

You will be redirected to:

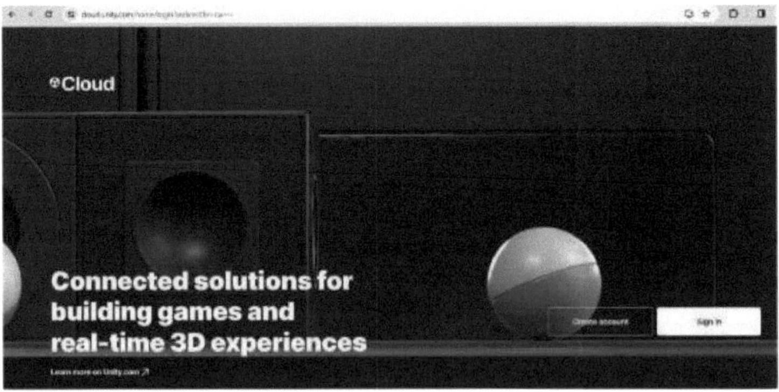

Select the field:

We personalise:

We will have a panel at the cloud level

Then to create a project we log into unity hub, select all the templates, and then we create a new project.

We select Microgame Karting

In this screen we see a tutorial of the construction of the game:
We can observe some stages of the construction of the video game
Go to File, game settings and select WebGL
Having:

5. CONCLUSIONS:

- The first steps in Unity provide a fundamental introduction to game development and Unity's intuitive user interface.
- Through the creation of scenes and the manipulation of objects, developers can begin to build environments and define interactions.
- C# scripts allow you to define the behaviour of objects, including movement and interaction with other game elements.
- Configuring the controls is essential to provide a comfortable gaming experience for players. Furthermore, understanding the concepts of physics and collisions in Unity is crucial to create interactive and realistic games.
- In short, these first steps lay the groundwork for more complex and ambitious projects in Unity, offering the opportunity to create a variety of impressive and creative games with practice and dedication.

6. BIBLIOGRAPH(A:

Bucheli Waters, L. F. (2023a). Mastering Object Oriented Programming with Java in NetBeans. Our Knowledge Publishing.

- Aguas Bucheli, L. F. (2023b). Hands on: Lab Practices in Data Structures. Our Knowledge Publishing.
- Aguas, L., Recalde, H., Toasa, R., & Salazar, E. (2023). Design of a video game applying a layered architecture based on the unity framework. In Lecture Notes in Networks and Systems (pp. 535-550). Springer International

Publishing.

- Aguas, L. (2023). 3D modelling of the OKmpico Atahualpa Stadium using the hypermedia design methodology. Proquest.com. https://proquest.com/openview/dc587bdc2c3025c64fc7f9f423a34c38/ 1?pq-origsite=gscholar&cbl=1006393
- Aguas, L., Suarez, L., Coral, R., & Machay, B. (2023). 3D modelling of freedom summit for virtual environments. In Trends in Artificial Intelligence and Computer Engineering (pp. 548-560). Springer Nature Switzerland.

1. **TOPIC:** Pong game in Unity
2. **OBJECTIVES:**

- To familiarise with Unity as a comprehensive platform for the development of real-time 3D interactive content.
- Analyse the Unity interface.

3. **INTRODUCTION:**

The game Pong is one of the earliest arcade video games and can be considered the forerunner of modern video games. It is a simple two-dimensional tennis simulation, where players control paddles that move vertically at the ends of the screen to hit a ball to the opponent's side. The goal is to avoid the ball going past their side while trying to get the ball past the opponent's side. It was created in the 1970s and achieved great popularity, marking the beginning of the video game industry.
The game Pong in Unity is a digital recreation of the classic ping pong game. Two players control rods (paddles) on the screen, with the goal of hitting a ball and sending it to the opponent's side without the opponent being able to return it. The player scores a point when the ball passes the opponent's side and is not returned.
To implement it in Unity, we set up a 2D project, create the sprites for the ball and the paddles, add physics with RigidBody2D and BoxCollider2D, and program the movement and logic of the game with C# scripts.

4. **DEVELOPMENT:**

We open Unity and create a new project.
Then in MainCamera, select the background in black colour
We create a new one with the name Ball
We place in Inspector
We duplicate
We rename it Paddle1, and in inspector we place
We duplicate Paddle1, and place in the name Paddle2 and in inspector we place
We duplicate Paddle2 and name it TopWall and in Inspector we place
Duplicate the TopWall and name it BottonWall and in Inspector we place
We duplicate the BottonWall and name it Goal1 and in Inspector we place
We duplicate Goal1 and name it Goal2 and in Inspector we place
We duplicate Goal2 and name it MiddleLine and in Inspector we place
We create a script with the name Ball
We place the following code by double clicking on it

```
using System.Collections;
using System.Collections.Generic;
using UnityEngine;
public class Ball : MonoBehaviour
{
[SerializeField] private float initialVelocity = 4f;
[SerializeField] private float VelocityMultiplier = 1.1f;
private Rigidbody2D ballRb;
void Start()
{
ballRb = GetComponent<Rigidbody2D>();
Launch();
}
private void Launch()
{
float xVelocity = Random.Range(0,2) ==0 ? 1 : -1;
float yVelocity = Random.Range(0,2) == 0 ? 1 : -1;
ballRb.velocity=new Vector2(xVelocity,yVelocity)*initialVelocity;
}
private void onCollisionEnter2D(Collision2D collision)
{
ballRb.velocity*= VelocityMultiplier;
}
void Update()
{
}
}
```

We create another script with the name Paddle

Then we double click and place the following code

```
using System.Collections;
using System.Collections.Generic;
using UnityEngine;
public class Paddle : MonoBehaviour
{
[SerializeField] private float speed = 7f;
[SerializeField] private bool isPaddle1;
private float yBound = 3.75f;
```

```
void Update()
{
float movement;
if(isPaddle1)
{
movement = Input.GetAxisRaw("Vertical");
}
else
{
movement = Input.GetAxisRaw("Vertical2");
}
Vector2 paddlePosition = transform.position;
paddlePosition.y = Mathf.Clamp(paddlePosition.y + movement * speed *
Time.deltaTime, -yBound, yBound);
transform.position = paddlePosition;
}
}
```

We place the name of Bouncy and in Inspector we place
Select Ball and add a component Box Collider 2D and RigidBody2D
In Box Collider 2D and RigidBody2D we place Bouncy material and the RigidBody2D in Collision Detection that is Continuous.
We add another component.
In Paddle1 and Paddle2 we add a Box Collider 2D and RigidBody2D component.
Drag the Paddle script **to** Paddle1 and Paddle2.
Then in Paddle1
Select TopWall and add a Box Collider 2D component.
Select BottonWall and add a Box Collider 2D component.
Select Goall and add a Box Collider 2D component.
Select Goal2 and add a Box Collider 2D component.
In Goal1 and Goal2, we mark
In Paddle1 and Paddle2, we select Kinematic
Let's go to:

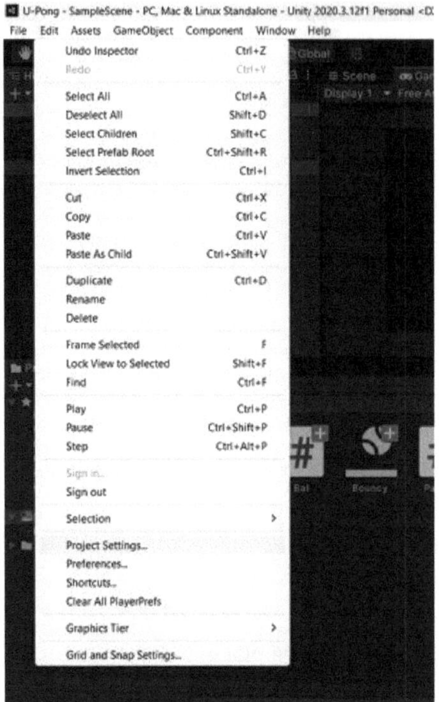

5. VISUALISATION

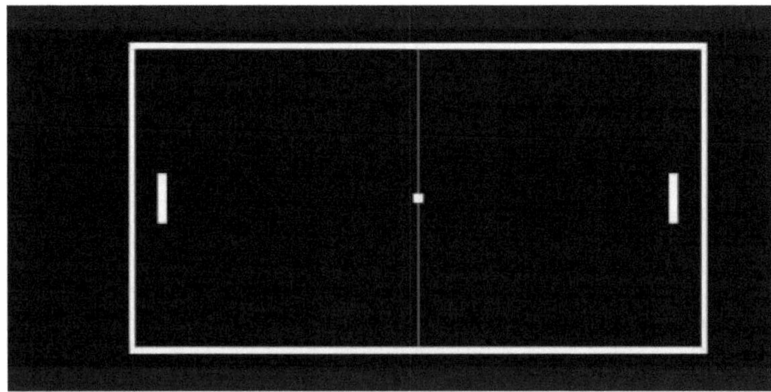

6. CONCLUSIONS:

Implementing the Pong game in Unity can offer several interesting conclusions:

- **Learning Physics and Controls:** Implementing Pong in Unity is an excellent opportunity to learn about Unity's physics system, especially how objects interact and react to collisions. It also allows you to practice creating responsive and smooth user controls.

- **Game Programming Fundamentals:** Pong is an ideal project for beginners, as it covers many of the basic concepts of game programming, such as game loops, game states, and event handling.
- **Optimisation and Performance:** Although Pong is a simple game, optimising performance and ensuring that it runs smoothly can be a good exercise. This includes efficient memory management and code optimisation.
- **Creativity and Expansion:** Once the basic version of Pong is mastered, there is ample room for creativity. Additional features such as power-ups, different game modes, and improved graphics can be added, which can lead to a deeper understanding of Unity and game development in general.

7. BIBLIOGRAPH(A:

- Bucheli Waters, L. F. (2023a). Mastering Object Oriented Programming with Java in NetBeans. Our Knowledge Publishing.
- Aguas Bucheli, L. F. (2023b). Hands on: Lab Practices in Data Structures. Our Knowledge Publishing.
- Aguas, L., Recalde, H., Toasa, R., & Salazar, E. (2023). Design of a video game applying a layered architecture based on the unity framework. In Lecture Notes in Networks and Systems (pp. 535-550). Springer International Publishing.
- Aguas, L. (2023). 3D modelling of the OKmpico Atahualpa Stadium using the hypermedia design methodology. Proquest.com. https://proquest.com/openview/dc587bdc2c3025c64fc7f9f423a34c38/Ppq-origsite=gscholar&cbl=1006393
- Aguas, L., Suarez, L., Coral, R., & Machay, B. (2023). 3D modelling of freedom summit for virtual environments. In Trends in Artificial Intelligence and Computer Engineering (pp. 548-560). Springer Nature Switzerland.

1. **TOPIC:** Character Insertion
2. **OBJECTIVES:**

- To familiarise with Unity as a comprehensive platform for the development of real-time 3D interactive content.
- Analyse the Unity interface.

3. **INTRODUCTION:**

3D components in Unity are those that are used to manipulate objects and create visual effects in a three-dimensional environment. Here are some of the most common 3D components in Unity:

- **Transform:** This is the basic component that all GameObjects have in Unity. It controls the position, rotation and scale of the object in 3D space.
- **Mesh Renderer:** This component allows an object to be rendered in the Unity scene. It defines how the object is rendered, including materials, textures and lighting.
- **Mesh Filter:** This component is used to specify the geometric shape of a 3D object. It is used to attach meshes to GameObjects to define their shape.
- **Collider: Colliders** are components that are used to detect collisions between objects in the scene. They can be simple geometric shapes such as cubes, spheres or capsules, or they can be more complex meshes.
- **Rigidbody:** This component is used to simulate physics in GameObjects. Adding a Rigidbody allows an object to respond to gravity, force and other physical forces.
- **Animator:** This component is used to control animations in a GameObject. It allows the playback of animations created in Unity, such as walking, running, jumping, etc.
- **Light:** Light components are used to illuminate the scene. They can be directional, spot, area or spot lights, and affect how objects look in the scene.
- **Audio Source:** This component is used to play sounds in a GameObject. It can be used to add sound effects to objects in the scene.
- **Particle System:** This component is used to create and control particle effects, such as fire, smoke, sparks, etc. It is very useful for creating complex and dynamic visual effects.

Export the object from Blender:

- Open your project in Blender.
- Select the object you wish to export.
- Go to the "File" menu and select "Export".

- Choose your preferred file format. The FBX format is commonly used to export 3D models to Unity. You can also use other formats such as OBJ.
- Configure the export options according to your needs and save the file in an accessible location.

Import the object into Unity:

- In Unity, open your project.
- Navigate to the folder where you saved the file exported from Blender.
- Right click on the "Assets" folder in the "Project" window and select "Import New Asset".
- Select the FBX file (or your chosen format) that you exported from Blender and click "Import".
- Unity will import the file and place it in your project.

Use the object in your scene:

- Once imported, you can drag and drop the object from the Project window into your scene in the Hierarchy or Scene window.
- You can also instantiate the object in your scene from the "GameObject" menu in the toolbar, selecting "3D Object" and then the corresponding object.

Adjust the configuration as necessary:

- Once the object is in your scene, you can adjust its position, rotation and scale using the manipulation tools in the scene.
- You can apply materials and textures to the object as needed.

4. DEVELOPMENT:

We open Unity and create a new project.
We download a component from:
https://mega.nz/folder/yAtUXAqK#QdQ4LJKWxPwhHsiAWhTgqg
Click on the Assets area, having:
We import new Asset
Drag the component to the scene
Expand the folder of the added object
We click on the character and place in inspector
We create a new script
We place the following code

```
using UnityEngine;
public class AnimationPersonage : MonoBehaviour
{
private Animator animator; // Reference to Animator component
```

```
private Rigidbody rb; // Reference to Rigidbody component
public float velocityMovement = 5.0f; // Character movement speed
void Start()
{
// Obtain references to the necessary components
animator = GetComponent<Animator>();
rb = GetComponent<Rigidbody>();
}
void Update()
{
// Get the horizontal and vertical movement input
float movementHorizontal = Input.GetAxis("Horizontal");
float movementVertical = Input.GetAxis("Vertical");
// Update Rigidbody speed based on input
Vector3 motion = new Vector3(motionHorizontal, 0.0f, motionVertical);
rb.velocity = movement * velocityMovement;
// If the character is moving, activate the walk animation.
if (movement.magnitude > 0)
{
animator.SetBool("Walking", true);
}
else // If not moving, disable walking animation
{
animator.SetBool("Walking", false);
}
}
}
```

Having

Drag the script to the 3d object

Then we add component RigidBody

Then we add component Animator

And if we click on the play button, we can see that it moves vertically.

We create another script

```
using UnityEngine;
public class WalkingPersonage : MonoBehaviour
{
public float velocity = 2.0f; // Character movement speed
```

```
public float velocityRotation = 100.0f; // Character rotation speed
private Animator animator; // Reference to Animator component
void Start()
{
// Get the Animator component from the GameObject animator =
GetComponent<Animator>();
}
void Update()
{
// Get horizontal and vertical movement input
float movementHorizontal = Input.GetAxis("Horizontal");
float movementVertical = Input.GetAxis("Vertical");
// Calculate the motion vector based on the player's input
Vector3 motion = new Vector3(motionHorizontal, 0.0f, motionVertical);
// Normalise the motion vector to maintain a constant velocity if the player is
moving diagonally.
movement = movement.normalized;
// Rotate the character to the direction of movement
if (movement != Vector3.zero)
{
Quaternion rotation = Quaternion.LookRotation(motion);
transform.rotation = Quaternion.RotateTowards(transform.rotation, rotation,
rotationSpeed * Time.deltaTime);
}
// Apply movement to the character
transform.Translate(motion * velocity * Time.deltaTime, Space.World);
// Update the animation parameter "Speed" in the Animator
float velocityAnimation = movement.magnitude;
animator.SetFloat("Speed", speedAnimation);
}
}
```

Having:

We create an animator controller

Under the name CharacterController

We select the previous component and create a new state with the name of walk

We create an avatarmask

We create new animation
Then click on play.

5. CONCLUSIONS:

- Animating objects in Unity 3D is an essential part of the game and interactive application development process. Unity offers a wide range of tools and techniques for animating objects, allowing developers to create a variety of visual effects and behaviours for their games and applications.
- Unity provides a set of integrated animation tools, such as the Mecanim system for humanoid characters and the particle system for visual effects, which allow developers to create complex animations efficiently.
- Objects in Unity can be animated programmatically using C# or JavaScript scripts. This gives developers precise control over animation behaviour and the ability to create interactive and dynamic animations.
- It is important to optimise animations in Unity to ensure optimal performance, especially on mobile devices and low-powered platforms. This may include the use of techniques such as using event-driven animations and reducing the number of vertices in animated meshes to minimise computational cost.
- In short, animating objects in Unity 3D is a fundamental task for creating engaging and interactive visual experiences in games and applications. With the right tools and techniques, developers can achieve a wide variety of visual effects and animated behaviours in their projects.

6. BIBLIOGRAPH(A:

- Bucheli Waters, L. F. (2023a). Mastering Object Oriented Programming with Java in NetBeans. Our Knowledge Publishing.
- Aguas Bucheli, L. F. (2023b). Hands on: Lab Practices in Data Structures. Our Knowledge Publishing.
- Aguas, L., Recalde, H., Toasa, R., & Salazar, E. (2023). Design of a video game applying a layered architecture based on the unity framework. In Lecture Notes in Networks and Systems (pp. 535-550). Springer International Publishing.
- Aguas, L. (2023). 3D modelling of the OKmpico Atahualpa Stadium using the hypermedia design methodology. Proquest.com. https://proquest.com/openview/dc587bdc2c3025c64fc7f9f423a34c38/Ppq-origsite=gscholar&cbl=1006393
- Aguas, L., Suarez, L., Coral, R., & Machay, B. (2023). 3D modelling of

freedom summit for virtual environments. In Trends in Artificial Intelligence and Computer Engineering (pp. 548-560). Springer Nature Switzerland.

1. **TOPIC:** Asset Store
2. **OBJECTIVES:**

- To familiarise with Unity as a comprehensive platform for the development of real-time 3D interactive content.
- Analyse the Unity interface.

3. **INTRODUCTION:**

3D components in Unity are those used to manipulate objects and create visual effects in a three-dimensional environment. Here are some of the most common 3D components in Unity:

- **Transform:** This is the basic component that all GameObjects have in Unity. It controls the position, rotation and scale of the object in 3D space.
- **Mesh Renderer:** This component allows an object to be rendered in the Unity scene. It defines how the object is rendered, including materials, textures and lighting.
- **Mesh Filter:** This component is used to specify the geometric shape of a 3D object. It is used to attach meshes to GameObjects to define their shape.
- **Collider: Colliders** are components that are used to detect collisions between objects in the scene. They can be simple geometric shapes such as cubes, spheres or capsules, or they can be more complex meshes.
- **Rigidbody:** This component is used to simulate physics in GameObjects. Adding a Rigidbody allows an object to respond to gravity, force and other physical forces.
- **Animator:** This component is used to control animations in a GameObject. It allows the playback of animations created in Unity, such as walking, running, jumping, etc.
- **Light:** Light components are used to illuminate the scene. They can be directional, spot, area or spot lights, and affect how objects look in the scene.
- **Audio Source:** This component is used to play sounds in a GameObject. It can be used to add sound effects to objects in the scene.
- **Particle System:** This component is used to create and control particle effects, such as fire, smoke, sparks, etc. It is very useful for creating complex and dynamic visual effects.

Export the object from Blender:

- Open your project in Blender.

- Select the object you wish to export.
- *Go* to the "File" menu and select "Export".
- Choose your preferred file format. The FBX format is commonly used to export 3D models to Unity. You can also use other formats such as OBJ.
- Configure the export options according to your needs and save the file in an accessible location.

Import the object into Unity:

- In Unity, open your project.
- Navigate to the folder where you saved the file exported from Blender.
- Right click on the "Assets" folder in the "Project" window and select "Import New Asset".
- Select the FBX file (or your chosen format) that you exported from Blender and click "Import".
- Unity will import the file and place it in your project.

Use the object in your scene:

- Once imported, you can drag and drop the object from the Project window into your scene in the Hierarchy or Scene window.
- You can also instantiate the object in your scene from the "GameObject" menu in the toolbar, selecting "3D Object" and then the corresponding object.

Adjust the settings as necessary:

- Once the object is in your scene, you can adjust its position, rotation and scale using the manipulation tools in the scene.
- You can apply materials and textures to the object as needed.

4. DEVELOPMENT:

Log in to assetstore.unity.com

Let's go to filters

Click on Add to My Assets.

You will be asked to create an account

Then click on Add to My Assets again.

Then Open in Unity

Then we create a new 2d project, with the character name

We go to Package Manager

Click on Download

Then click on import

We select all

Then under Import

We enter:
https://assetstore.unity.com/packages/2d/characters/simple-2d-platformer-assets-pack-188518
We are looking for this Asset

Anadimos

Download and Import
Then we rename the next one with the name Player
Having
We add a BoxCollider2D
We add a Rigibody2D

We select Continuous
We give Freeze on Z
We create a script with the name Movement

```
using System.Collections;
using System.Collections.Generic;
using UnityEngine;
public class MovementPlayer : MonoBehaviour
{
private Rigidbody2D rb2D;
```

```
[Header("Movement")]
private float float movementHorizontal = 0f;
[SerializeField] private float velocityOfMovement;
[SerializeField] private float smoothedFromMotion;
private Vector3 velocity=Vector3.zero;
private bool lookingRight = true;
private void Start()
{
rb2D = GetComponent<Rigidbody2D>();
private void Update() {
movementHorizontal = Input.GetAxisRaw("Horizontal") movementSpeed;
} private void FixedUpdate() {
//Move
Move(horizontalMovement * Time.fixedDeltaTime);
private void Mover(float mover) {
Vector3 velocityTarget= new Vector2(move,rb2D.velocity.y);
rb2D.velocity=Vector3.SmoothDamp(rb2D.velocity,velocityTarget,velocityref,S
moothMotionSmoothing);
if(move > 0 && !lookingRight)
{
//Turn
Rotate();
}
else if(moving < 0 && lookingRight) {
//Turn
Rotate();
}
}
private void Rotate()
{
lookingRight = !lookingRight;
Vector3 scale = transform.localScale;
scale.x *= -1;
transform.localScale = scale;
}
}
```

We create an Empty

Rename as: FloorController

We include in the script

```
[Header("Jump")]
[SerializeField] private float forceFromJump;
[SerializeField] private LayerMask thatIsFloor;
[SerializeField] private Transform controllerFloor;
[SerializeField] private Vector3 dimensionsBox;
[SerializeField] private bool enSuelo;
private bool jump=false;
```

Then we make further changes, having: using System.Collections;

```
using System.Collections.Generic;
using UnityEngine;
public class MovementPlayer : MonoBehaviour
{
private Rigidbody2D rb2D;
[Header("Movement")]
private float float movementHorizontal = 0f;
[SerializeField] private float velocityOfMovement;
[SerializeField] private float smoothedFromMotion;
private Vector3 velocity=Vector3.zero;
private bool lookingRight = true;
[Header("Jump")]
[SerializeField] private float forceFromJump;
[SerializeField] private LayerMask thatIsFloor;
[SerializeField] private Transform controllerFloor;
[SerializeField] private Vector3 dimensionsBox;
[SerializeField] private bool enSuelo;
private bool jump=false;
private void Start()
{
rb2D = GetComponent<Rigidbody2D>();
}
private void Update()
{
movementHorizontal = Input.GetAxisRaw("Horizontal") * movementSpeed;
if (Input.GetButtonDown("Jump"))
{
```

```
jump = true;
}
}
private void FixedUpdate()
{
inFloor = Physics2D.OverlapBox(controllerFloor.position, boxDimensions, 0f,
whichFloor.position);
//Move
Move(horizontalMovement * Time.fixedDeltaTime,jump);
jump = false;
}
private void Mover(float move,bool jump)
{
Vector3 velocityTarget= new Vector2(move,rb2D.velocity.y);
rb2D.velocity=Vector3.SmoothDamp(rb2D.velocity,velocityTarget,velocityref,S
moothMotionSmoothing);
if(move > 0 && !lookingRight)
{
//Turn
Rotate();
}
else if(moving < 0 && lookingRight) {
//Turn
Rotate();
}
if(enSuelo && skip)
{
enSuelo = false;
rb2D.AddForce(new Vector2(0f, jumpForce));
}
}
private void Rotate()
{
lookingRight = !lookingRight;
Vector3 scale = transform.localScale;
scale.x *= -1;
transform.localScale = scale;
```

```
}
private void OnDrawGizmos()
{
Gizmos.colour = Color.yellow;
Gizmos.DrawWireCube(controllerFloor.position, BoxDimensions);
}
}
```

On the floor:

We go to Soil Controller

Press play

5. BIBLIOGRAPH^A:

- Bucheli Waters, L. F. (2023a). Mastering Object Oriented Programming with Java in NetBeans. Our Knowledge Publishing.
- Aguas Bucheli, L. F. (2023b). Hands on: Lab Practices in Data Structures. Our Knowledge Publishing.
- Aguas, L., Recalde, H., Toasa, R., & Salazar, E. (2023). Design of a video game applying a layered architecture based on the unity framework. In Lecture Notes in Networks and Systems (pp. 535-550). Springer International Publishing.
- Aguas, L. (2023). 3D modelling of the OKmpico Atahualpa Stadium using the hypermedia design methodology. Proquest.com. https://proquest.com/openview/dc587bdc2c3025c64fc7f9f423a34c38/Ppq-origsite=gscholar&cbl=1006393
- Aguas, L., Suarez, L., Coral, R., & Machay, B. (2023). 3D modelling of freedom summit for virtual environments. In Trends in Artificial Intelligence and Computer Engineering (pp. 548-560). Springer Nature Switzerland.

ADVANCED PRACTICES

Practice 1: Artificial Intelligence for NPCs

Theme

Implementation of artificial intelligence (AI) behaviours for non-player characters (NPCs).

Objectives

- Understand the concepts of artificial intelligence in games.
- Implement navigation routes and autonomous movement for NPCs.
- Apply behavioural techniques such as tracking, patrolling and evasion.

Development

1. **Introduction to AI in Unity**
   - Brief explanation of the basic concepts of AI in videogames.
   - Description of the navigation components and their configuration.
2. **Scenario Configuration**
   - Create a terrain suitable for the navigation of NPCs.
   - Define obstacles and navigable areas using the NavMesh system.
3. **Implementation of Autonomous Movements**
   - Add the NavMeshAgent component to the NPCs.
   - Set up patrol routes and points of interest.
   - Write scripts in C# to control tracking and evasion behaviours.
4. **Simulation and Testing**
   - Run the scene and observe the behaviours of the NPCs.
   - Adjust navigation and behavioural parameters to optimise AI.

BIBLIOGRAPH^A:

- Bucheli Waters, L. F. (2023a). Mastering Object Oriented Programming with Java in NetBeans. Our Knowledge Publishing.
- Aguas Bucheli, L. F. (2023b). Hands on: Lab Practices in Data Structures. Our Knowledge Publishing.
- Aguas, L., Recalde, H., Toasa, R., & Salazar, E. (2023). Design of a video game applying a layered architecture based on the unity framework. In Lecture Notes in Networks and Systems (pp. 535-550). Springer International Publishing.
- Aguas, L. (2023). 3D modelling of the OKmpico Atahualpa Stadium using the hypermedia design methodology. Proquest.com. https://proquest.com/openview/dc587bdc2c3025c64fc7f9f423a34c38/Ppq-origsite=gscholar&cbl=1006393
- Aguas, L., Suarez, L., Coral, R., & Machay, B. (2023). 3D modelling of freedom summit for virtual environments. In Trends in Artificial Intelligence and Computer Engineering (pp. 548-560). Springer Nature Switzerland.

Practice 2: Augmented Reality with AR Foundation

Theme

Development of augmented reality (AR) applications using AR Foundation.

Objectives

- To become familiar with the basic concepts of augmented reality.
- Setting up a Unity project for AR.

- Implement AR elements such as surface detection and anchoring of virtual objects.

Development

1. **Introduction to AR Foundation**

- Explanation of the basics of augmented reality.
- Installation and configuration of AR Foundation in Unity.

2. **Project configuration**

- Create a new Unity project configured for AR.
- Import and configure the AR Foundation package.

3. **Implementation of AR Functionalities**

- Configure the AR camera and the surface detection system.
- Adding scripts to anchor virtual objects in the real world.
- Implement basic interactions such as touching to place objects.

4. **Device Testing**

- Implement and test the application on compatible mobile devices.
- Make adjustments and optimisations based on the evidence.

BIBLIOGRAPH^A:

- Bucheli Waters, L. F. (2023a). Mastering Object Oriented Programming with Java in NetBeans. Our Knowledge Publishing.
- Aguas Bucheli, L. F. (2023b). Hands on: Lab Practices in Data Structures. Our Knowledge Publishing.
- Aguas, L., Recalde, H., Toasa, R., & Salazar, E. (2023). Design of a video game applying a layered architecture based on the unity framework. In Lecture Notes in Networks and Systems (pp. 535-550). Springer International Publishing.
- Aguas, L. (2023). 3D modelling of the OKmpico Atahualpa Stadium using the hypermedia design methodology. Proquest.com. https://proquest.com/openview/dc587bdc2c3025c64fc7f9f423a34c38/Ppq-origsite=gscholar&cbl=1006393
- Aguas, L., Suarez, L., Coral, R., & Machay, B. (2023). 3D modelling of freedom summit for virtual environments. In Trends in Artificial Intelligence and Computer Engineering (pp. 548-560). Springer Nature Switzerland.

Practice 3: Multiplayer in Lmea with Photon

Theme

Development of multiplayer games in Knea using Photon Unity Networking (PUN).

Objectives

- Understand the principles of multiplayer development.
- Configure a Unity project to use Photon.
- Implement multiplayer functionalities such as waiting rooms, object synchronisation and chat.

Development

1. **Introduction to Photon**
   - Explanation of the basic concepts of multiplayer development.
   - Description of Photon and its features.
2. **Project configuration**
   - Create a new Unity project and configure Photon.
   - Register a Photon application and obtain the necessary credentials.
3. **Implementation of Multiplayer Functionalities**
   - Create and join waiting rooms.
   - Synchronise the movement and state of objects between players.
   - Implement a basic chat system.
4. **Testing and Optimisation**
   - Run the application on multiple devices to test multiplayer functionality.
   - Make adjustments to optimise latency and synchronisation.

BIBLIOGRAPH^A:

- Bucheli Waters, L. F. (2023a). Mastering Object Oriented Programming with Java in NetBeans. Our Knowledge Publishing.
- Aguas Bucheli, L. F. (2023b). Hands on: Lab Practices in Data Structures. Our Knowledge Publishing.
- Aguas, L., Recalde, H., Toasa, R., & Salazar, E. (2023). Design of a video game applying a layered architecture based on the unity framework. In Lecture Notes in Networks and Systems (pp. 535-550). Springer International Publishing.
- Aguas, L. (2023). 3D modelling of the OKmpico Atahualpa Stadium using the hypermedia design methodology. Proquest.com. https://proquest.com/openview/dc587bdc2c3025c64fc7f9f423a34c38/Ppq-origsite=gscholar&cbl=1006393
- Aguas, L., Suarez, L., Coral, R., & Machay, B. (2023). 3D modelling of freedom summit for virtual environments. In Trends in Artificial Intelligence and Computer Engineering (pp. 548-560). Springer Nature Switzerland.

Practice 4: Custom Shaders with Shader Graph

Theme

Creation and implementation of custom shaders using Shader Graph.

Objectives

- Understand the basic concepts of shaders and how they work.
- Use Shader Graph to create shaders visually.
- Implement and apply custom shaders on materials and objects.

Development

1. **Introduction to Shaders**

  - Explanation of what shaders are and their importance in game graphics.
  - Description of the types of shaders and their basic operation.

2. **Project configuration**

  - Create a new Unity project and set up the environment for Shader Graph.
  - Install and configure the Shader Graph package.

3. **Creation of Custom Shaders**

  - Use Shader Graph to create visual effects such as water, fire and fog.
  - Configure nodes and properties to adjust shader behaviour.
  - Apply shaders to materials and objects in the scene.

4. **Testing and Adjustments**

  - Visualise and test the effects in different lighting and camera conditions.
  - Make adjustments to nodes and properties to improve visual effects.

BIBLIOGRAPH^A:

- Bucheli Waters, L. F. (2023a). Mastering Object Oriented Programming with Java in NetBeans. Our Knowledge Publishing.
- Aguas Bucheli, L. F. (2023b). Hands on: Lab Practices in Data Structures. Our Knowledge Publishing.
- Aguas, L., Recalde, H., Toasa, R., & Salazar, E. (2023). Design of a video game applying a layered architecture based on the unity framework. In Lecture Notes in Networks and Systems (pp. 535-550). Springer International Publishing.
- Aguas, L. (2023). 3D modelling of the OKmpico Atahualpa Stadium using the hypermedia design methodology. Proquest.com. https://proquest.com/openview/dc587bdc2c3025c64fc7f9f423a34c38/Ppq-origsite=gscholar&cbl=1006393
- Aguas, L., Suarez, L., Coral, R., & Machay, B. (2023). 3D modelling of freedom summit for virtual environments. In Trends in Artificial Intelligence and Computer Engineering (pp. 548-560). Springer Nature Switzerland.

Printed by Books on Demand GmbH, Norderstedt / Germany